FLUID MECHANICS:
A Fairy Tale

SARAH ALLEN

Cover design by Vik N. Charlie.

Illustrated by Kimberly Delain.

Except for chapter page illustration and stars.

Book design by Sue Balcer.

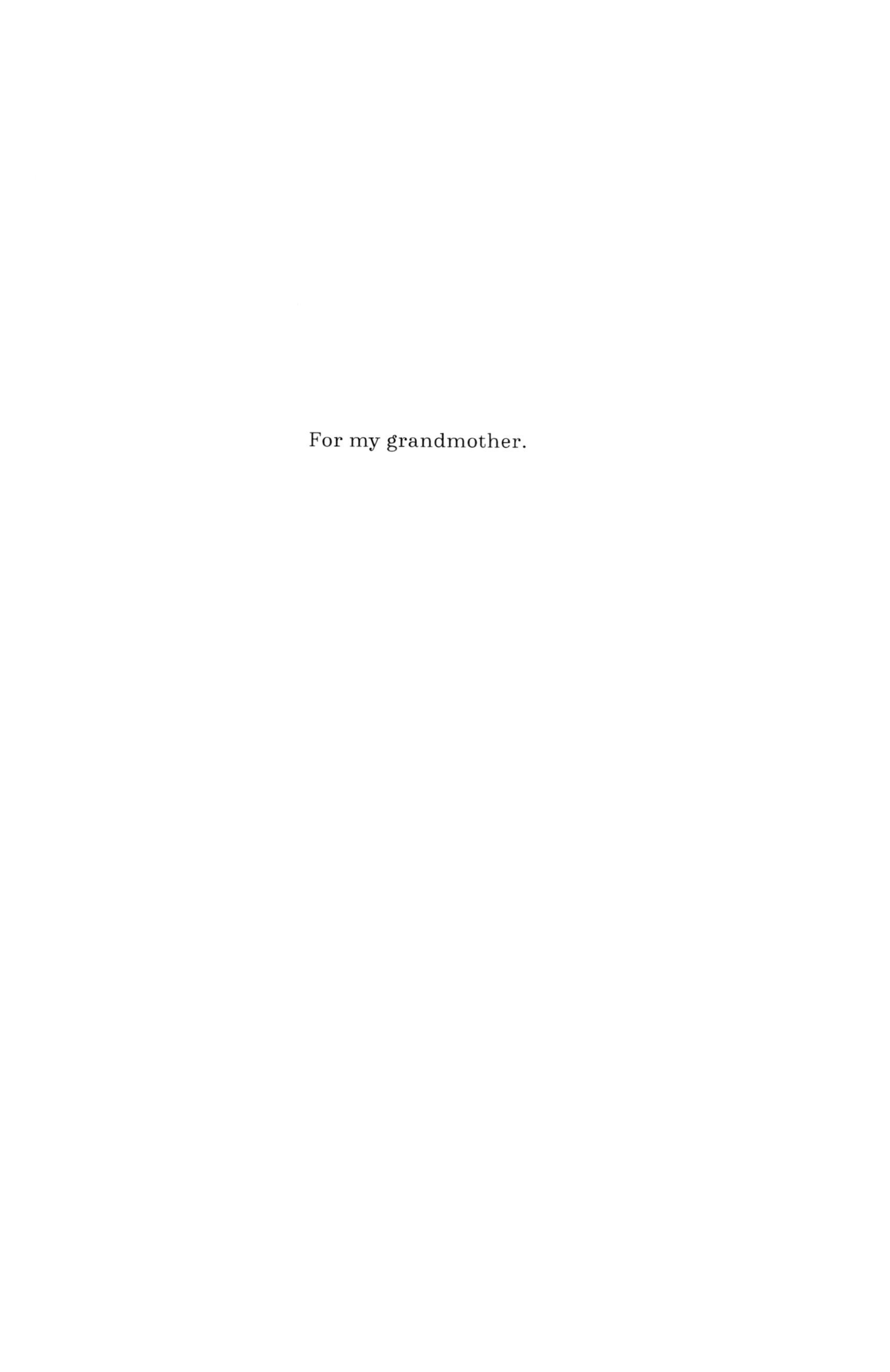

For my grandmother.

FLUID MECHANICS: A Fairy Tale

ONCE upon a time there lived a princess in a castle. Her mother had died when she was a baby, and she was raised by her father, King Alexander.

The king didn't have much time for Princess Elise. He awoke three hours before sunrise every day and went to bed three hours after sunset. He worked tirelessly to make the kingdom into the kind of place Elise's mother had always wanted it to be: prosperous, safe, and full of happy citizens.

Elise learned early on that to spend time with her father, she had to work. From the age of three she would sit on the floor next to his desk, reading books on economics, supply chains, and military readiness, carefully sounding out words like flotilla and echelon.

At first, she would only sit silently beside his throne while he discussed issues with his advisors, but it wasn't long before she began to speak up, first to ask questions, and then to make suggestions. As she grew, she gained a reputation for wisdom, intelligence, and the same work ethic that had earned her father so much respect.

Elise's best friend was a student named Edmond. They grew up attending classes together and would spend hours arguing political theory and exploring the castle.

Over the years, as Elise grew into a young adult, she slowly took over more and more of the duties of running the kingdom. For a long time, all was well, until one winter the king became deathly ill. He called Elise to his bedside, taking her hand and looking at her seriously.

He cleared his throat. "You must marry before I die," he said at last.

Of all the things Elise had expected, this was the furthest from her mind. Completely confused, she asked him to explain, but he only shook his head, saying that it was important. He was barely able to look her in the eye now.

"But why?" Elise asked. He'd never mentioned anything like this before. "Am I not capable of running the kingdom on my own?" It was her worst fear. Not that she was incapable, but that her father might think her so.

A deep sadness came over the king, and he took her hand, squeezing it tightly. "Of course not. You are more than capable." His eyes darted away. She could see there was something he wasn't telling her.

"What is it? If I'm capable, then why must I marry?" She racked her brain, going over every law, ordinance, treaty and sub-treaty she could think of. There was nothing. No alliance

that needed to be made, although perhaps there was some benefit she wasn't considering.

Her father shook his head, releasing her hand to cover a cough. The doctors swarmed around his bedside, and Elise was ushered from the room.

Two suitors arrived at the castle the very next day.

Full of worry for her father, and confusion at this strange turn of events, Elise agreed to meet them, but she had little intention of marrying. In the back of her mind, she'd always assumed she and Edmond would marry eventually. But it was still a far-off thought.

The first suitor was tall, dark-haired, and handsome and gave her a golden dress trimmed with rubies. The second was tall, fair-haired, and handsome and gave her a silver dress decorated with spirals of emeralds.

Out of politeness, and in deference to her ill father, the princess had dinner with each of the princes in turn.

On the first night, she ate with the first prince, who gave her small golden treasures and told her of his vast gold mines in the East.

On the second night, she had dinner with the second prince, who gave her jewels and silver and told her of his silver mines in the West.

On the third night, Elise was on the verge of going to see her father again to tell him she must decline the proposals,

when Edmond arrived, blushing, in the throne room. He was carrying a large package, wrapped in burlap. With the gold and silver princes and all the courtiers looking on, he knelt in the center of the room, bowing his head.

"I'm sorry, Princess. I had to write to my brothers for a proper gift." He unwrapped the package and held up a strange green dress, made of discs of jade sewn together like fish scales.

As she had with the other two, she had dinner with Edmond that night. He asked how she was doing, and she told him more of her father's strange demand. Together, they talked it over, trying to figure out why the king would have suggested something like this, but nothing new occurred to them.

Toward the end of their dinner, Edmond tried awkwardly to propose, but Elise stopped him. She thanked him, but said it wasn't the right time. She needed to understand her father's request, and she didn't want to be rushed. Edmond let out a sigh of relief, nodding, and they moved on to other subjects. She asked about his brothers (Archimedes and Bernoulli), who were powerful wizards living in the North, on the edge of the great sea.

They were well, Edmond said, although they had gotten reports from sailors of strange flashes of light far out to sea.

The next morning, Elise visited her father again and told him she wasn't ready to marry.

"But what about the prince with the gold?" the king asked.

“Is there something I should know about him?” she asked, but the king hedged, shaking his head and twisting his blankets nervously in his hands.

Slowly, the king recovered, and Elise gradually took on running the kingdom. Many years passed, and he never brought up his strange demand again. Until one day his health began to fail again. Sensing his death was near, he called her to his bedside. He said needed to tell her a story.

“When I was young,” he began, pausing to catch his breath, “I was not a king. I was just a boy living in a village. One day a beautiful princess came riding through town, and I was entranced. I wanted to meet her. My parents and brothers urged me to forget about her. There was nothing I could do, they said. But I couldn’t give up. I had to meet her. I set out for the capital, intent on making my fortune and at the very least hoping to talk to her.

“On the road one night I met a traveler who was cold and hungry, and I offered to let him share my fire and my meager supply of food. He was grateful, and when he had eaten, he revealed that he was a powerful wizard and that, because of my kindness, he was willing to make a bargain with me. He would help me become king, but only if I would agree to make him king after me by letting him marry my daughter. But what if I have a son? I asked him. He said that I would have only one daughter. I agreed.”

The king looked down at his blankets. "I am ashamed to say that I never truly thought I would have to honor our agreement. Part of me didn't believe he could make me king, but really I just couldn't imagine that the future would ever really come. And . . . I never considered how it would affect you." He looked at her for a moment, then away again.

"He fulfilled his part of the bargain. I became a wildly successful merchant. I met your mother and we fell in love.

"I wish that I had made my fortune on my own, now. When I first became ill, I knew it was time to fulfill my promise to him. Soon after, he appeared. He demanded I force you to marry him. I could not bring myself to do that so I told him we must go through a formal ceremony; that you must appear to choose him. But I didn't tell you. I hoped that you would simply pick him and that everything would be all right."

"The golden prince?" she asked.

"Yes. He was the golden prince. I should have told you from the beginning that you must marry him. When I told him you refused to marry anyone, and that I wouldn't force you, he flew into a rage and disappeared. I haven't heard from him since."

Elise's stomach went cold. It felt like the stone tiles were shifting under her feet. It made sense now, why her father had worked so hard. He'd wanted to feel like he deserved to be king, when really, he had only bartered for it. Bartered for it with her freedom, without even telling her. And then refused to honor his promise.

"I'm sorry," the king said.

Elise nodded vaguely, but inside her chest was a storm of emotions.

Her father died that evening.

As she had learned from her father, Elise turned to work to distract herself. But even running the kingdom was full of conflict for her now.

In the back of her mind, she wondered where the wizard was, guessing it was only a matter of time of time before he returned.

Edmond was by her side every day after that, listening to her thoughts and helping where he could.

Time passed, and the wizard did not appear. Eventually, Elise decided she had to make her own choices, and that she wasn't responsible for her father's bargain. She was crowned queen, and she and Edmond were married.

Three days after the wedding, the wizard appeared.

Elise and Edmond were sitting in the empty throne room, talking. It was late at night, and rain lashed the stained-glass windows. Suddenly, the candle flames wavered and flared, and the golden prince appeared.

Elise jumped to her feet, glancing instinctively around, but the guards were outside. They were completely alone.

She cleared her throat, wishing she had thought of something to offer the wizard, or some way of appeasing him. Her mind raced, but the wizard spoke first.

"Thirty years ago, I gave your father everything he wanted, and in exchange I asked for one thing."

"I'm sorry—" Elise started to say, but he cut her off.

"I'm not interested in words. You're clearly just as trustworthy as your father."

Edmond stood, too, and started to speak, when the wizard whirled around, pointing a long finger at him.

Edmond's face went gray, and he collapsed.

"Until I get what was promised to me, everything you value will wither and die."

On the last word, the wizard vanished.

Elise ran desperately to Edmond. His eyes were closed, and his body was as still as if he were dead, but he was breathing. She summoned the court physician, who informed her that Edmond was apparently in a deep sleep.

She clenched her fists, angry at herself. Ignoring this problem was exactly what her father had done, and it had gone on long enough. She summoned her most trusted advisor.

"What do we know of this wizard?" she asked

"Your father told me he comes from a land far to the North. Farther north than the land of King Edmond's brothers. Across the ocean."

She lifted her chin. "I will find the wizard. I will travel to Edmond's brothers and get their help crossing the great sea. I will find him and put an end to this. You will rule the kingdom while I am gone." Her advisor blanched but nodded.

The queen took no one with her on her journey. She was a light and fast rider, and she didn't want to be encumbered.

It took her three weeks to reach the coastal city where Edmond's brothers lived. When she rode through the palace gates, she wore the jade dress that Edmond had given her, so

the brothers might recognize her. It was surprisingly light and strong. She felt like an armored fish.

The palace secretary brought her to the throne room and presented her to Edmond's two elder brothers.

Archimedes was short and fat and wore yellow robes embroidered with blue mathematical symbols. His hair was tidy and had already gone a little gray. Bernoulli was tall and thin, stately in grey robes with dark blue trim, his hair loose and long about his shoulders.

Archimedes set down the book he had been reading and strode over to her jovially.

"Little sister! To what do we owe this pleasure?" He looked around, confused. "Did you not bring Edmond? And where are your travelling companions?"

She looked at him seriously. "I come on important business, older brother." And she told him of the wizard and Edmond's sleep.

The brothers thought for a moment, then held a short, whispered conference. Bernoulli stepped forward. "We cannot travel with you, but we can each give you a gift to help you on your quest."

Archimedes stepped forward. "You will need to find the wizard. To help, I give you this." He handed her a bronze ball the size of her fist. "This ball always appears to be the same size, but it will become smaller as it gets closer to the wizard."

"What do you mean it always appears to be the same size? How do I tell if it's getting smaller, then?"

"To tell how big it is, you must use this." He gestured to a bucket of water sitting on the floor between them. Bending towards it he said, "Place your finger on the water line."

She knelt and touched her finger to the surface of the water. He dropped the ball into the bucket. It sank to the bottom, and the level of the water rose past her second knuckle. He continued his explanation.

"The larger the ball is, the more water it will displace, and the higher the water level in the bucket will be."

Elise nodded. That made sense.

"I invented it in the bathtub," Archimedes said seriously.

Elise wasn't quite sure what to say to that, so she picked the ball from the water, dried it off, and put it solemnly into her pocket, thanking him.

Bernoulli stepped forward.

"You will need a way to cross the great sea. I give you this boat."

He gestured to a small silver boat that three servants were carrying in. It looked barely large enough for one person and too heavy to float. In the front of the boat was an emerald basin filled with still silver water. On the water in the basin floated a tiny jade boat.

Bernoulli took out a small silver cup, which had three stoppered holes in its sides. One hole was near the rim, the second was lower down the side, and the third was almost at the bottom of the cup. He scooped water into the cup and placed it in the basin where it hovered. He pulled the cork from the first stopper, and a slow stream of water curved out from it, splashing down into the water below.

"If you place the boat on this stream, the boat will move at this speed."

He replaced the first cork and removed the second one. Water sprayed out more quickly this time.

"Here, where the water is deeper, it flows out more quickly. If you remove this cork and place the boat on this stream, the boat will move more quickly."

Again, he replaced the cork.

"The third and deepest cork, which I will not remove now, is the fastest. Using the faster speeds will drain the cup more quickly." He refilled the cup. "Use the speeds sparingly, because only I can refill the cup."

She nodded and thanked them both again. Archimedes asked her to stay and eat with them, but she shook her head, saying that she needed to find the wizard as soon as possible. Bernoulli quietly directed servants to take the boat through a side door, down some steps, through a tunnel, and out onto the rocky beach.

The silver boat sank a few inches but held perfectly steady as she stepped into it. When she was seated, Archimedes handed her an empty bucket. She reached over the side, scooped water into it, then reached into her pocket for the brass ball and dropped it in. The water level rose several inches.

"Good luck," Archimedes said, and they both smiled at her.

She moved to the front of the little boat, reached into the emerald basin, and uncorked the first cork. A small stream of water arced out of the cup. She plucked the jade boat from the

surface of the water and set it on the stream. It hung there disconcertingly, and the boat began to move forward at a smooth, steady pace, barely even rocking with the waves.

There was a thin tiller of white wood at the back of the boat, and when she moved it gently from side to side, the boat turned smoothly. Now she needed to know which direction to go. She pulled the tiller firmly towards her and the boat angled to the right, parallel to the shoreline. The water level in the bucket was still dropping, but more slowly now. Next, she gradually pushed the tiller farther and farther away from her, until she was travelling in exactly the opposite direction. As the boat turned, the water level lowered faster and faster, but then held steady for a moment and began to rise again. She adjusted her course until the water level was dropping its fastest, then sat back and looked around.

It was early evening now, a fog was gathering over the sea and moving in towards the black rocks and high cliffs offshore. Farther out, a bank of thick, bright fog waited for her. Fragments of it drifted inland, trailing up and over the cliffs. As the sun began to set, the water below her glowed pink, but within just a few minutes she had entered the white mass, and everything became dark and grey, eventually subsumed by the black of night.

The only light came from the basin, which glowed slightly, crowded by the fog. The water was calm at first, but as the night wore on, the small waves grew larger and larger, and soon the little boat was rocketing down one side of each giant wave and trudging up the other. She quickly lost all sense of direction and had to keep consulting the bucket to stay on course.

The air was cold and clammy, and she was soaked through with spray. The jade dress was warm, though. The stones seemed to have absorbed her body heat from the day before and now gave it back to her. It felt as if they had been sitting in the sun.

She did not sleep, and by morning she was exhausted, wishing she had let Archimedes convince her to stay for a meal because she wasn't sure how long her travelling food would last. But, as the sea calmed and the fog thinned and lifted in the morning sunlight, she began to feel a little better. Up ahead, a dark island appeared.

It was a tall rock protruding from the sea. Its sides were sheer cliffs, but she could just make out specks of green way at the top. A perilously narrow path scribbled up the side of the cliff.

At the base of the path was a small landing. She replaced the cork and tied the boat up with a grimy rope. Then she began the climb. It took her over an hour to reach the top, during which time she did not look down once. The last little bit was the worst; whoever had made the path seemed to have run out of interest by the end, and there was only a sheer rock wall with a few handholds. The handholds were dirty, and slippery with sand, gravel, and little bits of rock, some of which came off in her hand.

When she reached the top, she squirmed over it, dragging herself on her belly, glad of the protection of the stone dress. She lay gasping and shaking for a moment with her eyes closed, which was when she noticed how forceful the wind was way up there. She opened her eyes cautiously.

The top of the rock was small, no more than fifteen feet in diameter. It was not flat, but sloping, with jagged black rocks poking up randomly. A giant crack ran through the middle. The heather growing over the top was still damp with dew. A few scraggly trees hung on grimly, their branches twisted by the wind. In the middle, directly over the crack, stood a circle of thick rectangular stones. In the center of these was the wizard, watching her.

"I—" she croaked out, but he lifted a corner of his black cape, whirled, and disappeared. She spent several minutes looking for him, then several more waiting to see if he would come back. But he was nowhere to be found.

The climb down the side of the rock was worse than the climb up. She was more tired than before and there was no way to avoid looking down. To prepare herself mentally, she lay on the top and cried for a while. Slipping and sliding and taking frequent crying breaks, she made her way back down to the little boat. The sphere had expanded again, and the bucket appeared to be fuller than before.

Again, she uncorked the cup, placed the little jade boat upon the stream, and set off. She travelled the rest of the day and all through the night. The next morning, when the fog had cleared, she saw another island. This was a large, flat rock, low and close to the sea.

When she brought her little boat up to the stone dock, she saw smooth stone paths, square pools of water, green fruit trees, and moss-covered sculptures, all chipped and overgrown. The wind whistled emptily. Dark clouds gathered overhead, and it began to rain, slowly at first and then more heavily. In the center of the garden, she came to the remains of a large house. Only its stone foundations were left, and an immense fireplace. In the fireplace burned a roaring fire, and in front of that stood the wizard. She was ten feet away from him when he turned angrily and disappeared.

With a sigh, she sank down onto the hearth, feeling the heat from the fire warm against her face and hands.

How would she ever get near enough to him to convince him to change his mind? She sat up a little straighter, thinking of Edmond. She was making progress, she told herself. She'd already managed to find the wizard. She would just keep following him until he agreed to speak to her.

She pulled some bread from her bag, deciding to rest a few minutes before she continued, and her thoughts turned to her father. Her memories of him were now conflicted, tainted with anger. How could he have made such a bargain? And how could he have never told her? She'd always thought he was the hardest working, most honorable person she knew, but now she saw all his work in a different light. He hadn't earned being king. He'd made a deal with a wizard. And when it came time to honor that deal, he hadn't.

What did that mean about her own claim to the throne? Could she really continue being queen, now that she knew how her father had earned the title? Maybe she really should give the throne to the wizard. But she had no idea if he would be a good king. She couldn't do that to the people.

All these thoughts swirled through her mind as she sat by the fire, letting it warm up her wet clothing and hair. No answers came to her.

Again, she travelled all night and when the fog cleared the next morning, a third island appeared. This island was larger than the other two, and she could see nothing but sheer black cliffs all around it. The only way inside appeared to be a jagged fissure. It was about twenty feet wide, and water flowed out of it. The walls of the fissure shot straight up and were completely blanketed in ferns. Consulting her bucket, she pointed the boat up the fissure, noticing that the water level in the cup was significantly lower than when she had started, almost down to the level of the first hole.

For almost an hour, she piloted the boat up the canyon. It curved this way and that but stayed about the same width. Then, up ahead, she heard a rushing noise. Rounding a bend, the canyon abruptly narrowed. It was now only about ten feet wide, and the water was flowing twice as fast. It was too fast for her boat to make any progress against the current, so she replaced the first cork and removed the second. The little stream of water shot out faster than before, and her boat picked up speed, just barely able to fight against the current. The water level in the cup was now below the first hole; she would not be able to use the first speed anymore.

After another hour of travelling up the winding canyon, she rounded another bend and the walls narrowed even further. Now they were only five feet wide. She could reach out and touch them, and the water was yet again twice as fast as before. Again, her boat could make no progress against this current, so she removed the third and final cork. She was now slowly inching her way up the canyon, but she could see the water level in the cup dropping. It passed the second hole. As the water level lowered, the boat slowed down, just barely making progress.

The water was nearly gone when she rounded a bend and came to a small sandy beach. The last of the water trickled out as she leapt ashore, pulling the boat up past the waterline. She picked up the bucket and set off on a narrow dirt trail. Soon, she came to an intersection. There were no signs, but she checked the bucket, walking a little way down each of the paths, and chose the leftmost one. At first, the path sloped upwards through dense forest. Then she came to a small ridge and followed it a short way to the right before taking a path that switch-backed down the other side. The forest became darker, and the trees looked older. Here, where they were sheltered from the wind, they grew to an enormous thickness and a towering height. She walked quietly, noticing that she felt hungry again, but wasn't sure if any of the berries were edible. They were strange and unfamiliar to her.

At last, she came to a clearing. In the center of it was a small lake; a small cottage crouched on the shoreline. In the very center of the water was a very strange sight. A large

wooden ship floated there. The sails were ragged, and the wood looked a little green, but it floated. Elise could not see why it would be there. There was no way for it to reach the sea, no outlet from this tiny lake which it almost filled. It barely had room to move at all. On this ship stood the wizard, looking angrily down at her.

"Stop following me," the wizard said. "I told you. Your father and I had a bargain. Neither of you honored that bargain. Now you must pay the price."

"I know," she said. The wizard did have a point, even if what he had done was wrong. "And I'm sorry my father didn't fulfill his promise to you. What can I do to make up for the promise that was broken?"

The wizard considered this skeptically.

"I will offer you one chance. If you complete a task for me and make me a new promise, I will wake your husband from his sleep."

"What promise?"

"That on one day a year I will be king."

She considered this. "What would you do if you were king?"

The wizard looked surprised and not quite ready for the question. He opened his mouth and closed it again. Then he muttered something about maybe throwing a ball. She raised her eyebrows.

"What about the rest of the year?" she asked. "Would you be willing to use your magic to help the kingdom?"

He considered this. "Could I live in the castle? Maybe . . . in a very tall tower . . . that leans ominously to one side?"

One of her eyebrows raised even higher. Now that she was finally talking to him, the wizard wasn't at all like she'd expected. She wasn't sure it was wise or responsible to give the kingdom to this person, even if it was only one day a year, but he had been promised much more. And there might be huge benefits to having a wizard in court.

"I'm sure we could build something like that."

He gave a little excited hop, clicking his boots together.

"What is the task?" Elise asked.

He gestured to the trapped galleon below him. "You must free my boat and get it back out to the ocean."

She eyed it thoughtfully. "How did it get there?"

"A storm. I was shipwrecked here. I don't know how it happened, though, because I was knocked out."

"I'll try."

The lake was in a large, tree-filled valley, the sides of which were sheer rock rising a hundred feet into the air. Had the storm been so powerful it had lifted the boat over those walls? And why, if the wizard was so powerful, could he not free it himself?

She sat on a log to think. The wizard extended a ramp from his boat to the shore, crossed it, and went into his cabin. Maybe she could dismantle the ship and carry it, piece by piece, to the dock. Although, there was no way the ship would fit through the canyon anyway. Could she build some sort of rigging to lift the whole ship over the walls? Maybe she could secretly build a replica of the ship out on the ocean somewhere, then destroy this one. Except, she didn't know anything about shipbuilding. The ship bobbed in front of her, tauntingly.

A twig cracked and she turned to see the wizard walking towards her, carrying a tray laden with food.

He said to her, "I'll give you a sandwich if you give me one."

"Give you back one of your own sandwiches? Sure," she said.

He set to eating surprisingly ravenously. When they finished, he went back inside his cottage. She continued to contemplate the ship. She could probably use logs to roll it along. It wouldn't be impossible to get it to the base of one of the cliffs. Then maybe she could build some sort of lifting device with pulleys. She needed rope and an ax. She asked, and the wizard had both, but the rope looked old and frayed, and there wasn't enough of it, anyway. She would have to make rope.

She gathered grass all that afternoon. In the evening, she wove thick cords with it. For five days, she gathered grass during the day and wove it into rope in the evening. The wizard continued to bring her food and watch her progress without comment. He always offered her food with the condition that he have some, too. She wondered about this, and finally decided to ask.

"I was shipwrecked here as a young man," he explained. I was trapped here a year until a water fairy took pity on me and granted me magic powers to help me escape. But fairies don't like just giving gifts. They like their jokes. She gave me magic powers, but with the condition that I could never do anything for myself without bargaining for it with someone

else. Anything I wanted I had to trade another for it. I've been that way ever since."

Elise's eyes softened. "How do you survive? Can you not even eat or sleep without bargaining for it?"

"I've found ways to work around that. Bargains. But I've found that even without food I do not die." He sank into heavy silence, and she didn't press him further, contemplating what that must be like for him. She had a hard time imagining what it would be like to not be able to do anything for oneself. Horrible, probably.

When there was enough rope, the queen began to gather logs. She worked for four days setting up the transportation system. When everything was ready, she looped the ropes around trees, tying them securely to the ship. Then she positioned the logs and began to pull. The ship slowly rose out of the water, rolling along the logs she had painstakingly smoothed with the ax. It moved about ten feet when one of the ropes snapped and the ship slid all the way back into the water with a great splash.

Frustrated, Elise decided to take a break for the rest of the day and go for a swim. She took off the jade dress and wore just her cotton undergarments. The water was transparent turquoise and cold but refreshing. Usually, she liked to swim near the surface, but this time she decided to swim deeper and explore.

The underside of the ship was coated in barnacles, except for some long scratches in the sides. The grooves weren't deep enough to sink it, but it looked like it had been clawed by a giant animal. Elise was puzzled but couldn't look at them long before she had to return to the surface for breath. As she stood waist deep in the water, she licked her lips—salty. The water was not fresh water at all, but ocean water. This was surprising because as far as she could tell, the little lake was landlocked; it didn't even have a stream connecting it to the ocean. Come to think of it, the water was exceptionally clean for still water. No bugs or slime—it was a lot more like running water than stagnant. Maybe there was an underground stream. Diving down again, she examined the bottom closely, peering through the clear water. Her eyes stung a little from the salt, but it passed quickly.

The lake was much deeper than she'd expected. At the end of a short shelf, the bottom dropped off, plunging deeper than she could see. She came up for another breath and dove down deeper, quickly, as far as she could go. It kept going. She swam down fifteen feet before she reached the bottom. And here it didn't end but curved into a horizontal tunnel.

Desperate for breath, she shot back to the surface, gulped in air for a few minutes, and dove down again, this time following the tunnel a short way. It became very dark just a little way down the tunnel. Low on air again, she turned to head back to the surface, but realized she couldn't see a thing. She didn't know where the path to the surface was. Panicked, she began swimming as fast as she could, looking for a wall to guide her back. Then she saw a ray of light and swam frantically towards it, following it up and up and finally surfacing at last, gasping.

She was no longer in the lake.

The ocean waves crashed somewhere below her. The sun was hotter and brighter. She swam to the edge of a small

pool—like a very large tide pool—and climbed out onto some rocks to rest. She was on a shelf overlooking the ocean below. It was above the water level—although she didn't know if it was high tide or low at the moment. The wind blew little ripples across the water of the pool she had emerged from. A connecting tunnel. Somehow, the wizard's ship must have been forced through the tunnel by the storm.

She wrapped her arms around her knees and thought. How to get it back out, then? She watched the water speculatively for a few minutes and then smiled, remembering something she'd read as a young child. She dove back into the pool, swam carefully through the tunnel, and popped back out on the other side.

The logs she had collected were still sitting there, waiting for her to have a use for them. She immediately set-to with the ax.

The wizard came out later, watched her work for a few minutes, but said only, "New plan?"

She nodded, too out of breath from chopping to answer.

It took her seven days to build the little aqueduct—a system of hollowed-out logs to convey water away from the lake. Then came the tricky part. She had saved the largest log for last. This she hollowed out, partially with the axe, partially by burning it out. While it was burning, she used another log to carve a corkscrew. It took her a whole day to get the corkscrew shaped well enough that it would fit inside the hollowed-out log, but finally it worked. She fitted a handle to the top of it.

She placed one end of the screw in the water, the other end extended over her aqueduct. When she turned the screw, it lifted water out of the lake and deposited it in the logs so that it was carried over a small hill about a hundred feet away.

As the water was emptied, the boat sank deeper and deeper into the hole. The wizard seemed confused as to how this was helpful, but didn't say anything, just went inside his cottage to take a nap. When only a few feet of water remained at the bottom of a fifteen-foot rock hole, Elise loaded the screw into the ship. Then, she rigged up a system of pulleys to pull the ship through the tunnel.

Lastly, she repositioned her screw, only this time she thrust one end into the ocean. The top extended over the empty hole. Careful not to pour water on the boat, she refilled the tunnel, and the ship rose to the surface. She tied up the ship and swam back through the tunnel.

The wizard had apparently just come out of his cabin and was staring at the lake.

"Did you sink my boat?"

She grinned. "I have completed the task."

"What? How?" The wizard looked around, as if expecting to see the boat hidden behind a tree somewhere. "Where is it?"

"It's tied up on the outside of the island. I think it'll be ready to go as soon as it's high tide."

The wizard looked like he was having trouble believing his ears. Since it didn't look like he was going to say anything, she went on. "I agree to your terms. You can be king once a year for one day, as long as you promise to stay on as the court wizard and help the kingdom during the rest of the year."

The wizard grinned, spinning around, and clicking his heels again. "Fantastic. I'll gather my things."

The wizard had just a few things to pack, and then together they swam through the tunnel to get to the ship.

They made a quick trip across the ocean, docking in a small town where they sold the ship and purchased some new horses. From there it was only a few days' ride back to the castle.

The chief advisor was beside himself with relief when they arrived. He had hidden Edmond away in a back room, refused to let anyone see him, and had pretended to be acting on his orders.

"But I gave you authority when I left," Queen Elise said.

"Yes, but you were gone for weeks, Majesty. The people would never let me rule forever. It was starting to look like you were never coming back, and I needed a contingency plan."

"Which was?"

"I was prepared to tell the people that King Edmond was fine, hide him from everyone, have you declared dead, have him remarry, fake a new heir to the throne, and then fake Edmond's death to hide the evidence," he said grimly.

"Well ... I ..." As often happened with the advisor, the queen didn't really know what to say. "I'm glad you had a plan," she finished at last.

When Edmond awoke, he was overjoyed to see Elise safe and sound, and understandably skeptical of the addition of the wizard at court. But Elise was adamant that he had been promised the kingdom, and that giving him one day per year was a small price to pay to have a wizard around, even one who could be a little petulant when he didn't get what had been promised to him. After all, they already had a steward who could be

disturbingly practical; maybe a slightly too fair-minded wizard could be useful, too.

The king and queen were always careful of what they promised him, and for his part, the wizard proved to be a competent ruler, so much so that when they went on vacations, they often left him in charge, along with the royal advisor. There were a few memorable and unfortunate occasions where things did not go quite as smoothly as planned—the entire royal court once spent a few weeks as chipmunks—but for the most part the system worked well. And, by and large, everyone lived happily ever after.

The Physics

EDMOND'S two older brothers in the story, Archimedes and Bernoulli, were famous scientists in real life.

In fact, Archimedes is one of the most famous mathematicians of all time. He was already something of a celebrity when he was alive, although not for his mathematics. He was a Greek, born in 287 BCE, and lived in Syracuse, Italy, for most of his life. He was good friends with—or possibly a cousin of—the king, King Hieron II. Archimedes' obsession was pure mathematics. He was most proud of the theorems he created, his favorite being the relationship between the surface areas and volumes of cylinders and spheres, which he liked so much he had them inscribed on his tomb.

But what he was most famous for were his war machines. In his time, Syracuse was under attack from the Romans, and King Hieron II convinced Archimedes to create devices to fight them off.

It's not clear whether some of these accounts are exaggerated or apocryphal, but some accounts describe him using super-powered catapults, systems of mirrors to burn down attacking ships, and great claws that would grab whole ships and lift them out of the water.

But for Archimedes, all that was much, much less important than his mathematical theories. Most of what we know of him, and his work, comes from his books: On Plane Equilibriums (two books), Quadrature of the Parabola, On the Sphere and Cylinder (two books), On Spirals, On Conoids and Spheroids, On Floating Bodies (two books), Measurements of a Circle, and The Sandreckoner.

In The Sandreckoner, he attempts to calculate the number of grains of sand needed to fill the universe. To do this, he had to invent his own number system. The Greek system at the time used 27 letters and you could only use it to count up to 10,000 (10,000 was called a "myriad".) So, he invented a number system to count higher, and in the process discovered one of the fundamental laws of exponents ($x^a x^b = x^{a+b}$).

But King Hieron convinced him to build practical things, too. So, Archimedes invented the compound pulley, which can let a single person lift whole ships on their own. Possibly most famously in our time, he is the inventor of the field of hydrostatics, investigating how and why objects float, and how different liquids interact.

As part of this, he figured out a cool trick for measuring how much volume an object has (how much space it takes up).

The story goes that one time when King Hieron II gave a metal worker some gold to make a crown, he suspected the man might have mixed some silver into it and kept some of the gold for himself. He asked Archimedes if he could figure out if the crown was actually pure gold. This seemed like it would be possible, since they knew the density of gold (density being how much stuff is packed into an object. The denser an object is, the more stuff is crammed into a smaller space, and the heavier it is.) Except, the crown was so irregularly shaped that they couldn't figure out its volume.

The story goes that Archimedes was taking a bath (it was apparently hard to get Archimedes to leave his work and take a bath, and he would often still be writing mathematical formulas while bathing) when he noticed that when he got into the bath the water level in the tub rose. This made him realize that he could figure out the volume of an object by submerging it in water and measuring how much water it displaced.

For example, if you have a pool:

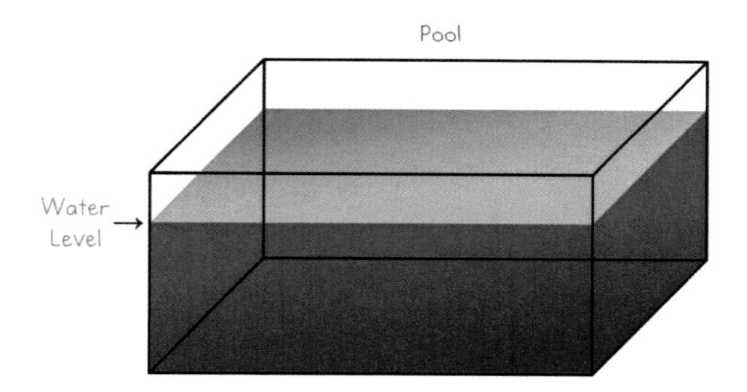

And you put a beach ball in it:

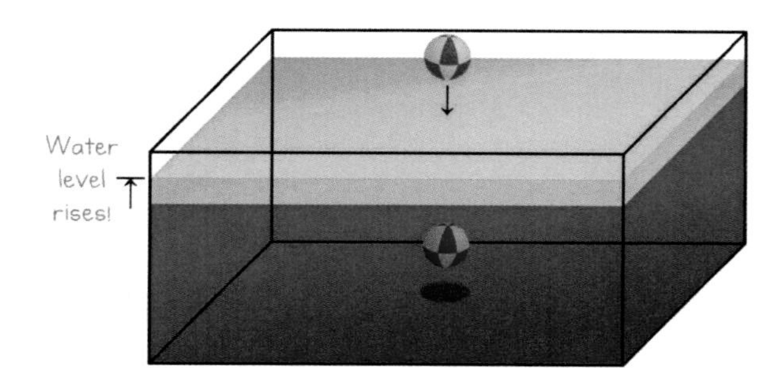

The water level rises by an amount that's exactly equal to the volume of the ball. The bigger the ball, the higher the level of the water.

That part of the story is probably true, but the next part is probably apocryphal. The story is that he leapt from the bath and ran naked down the streets shouting "Eureka!"

He measured the volume of the crown, and then, comparing it to its weight, found that the crown had indeed been diluted with silver.

The magical bronze ball Archimedes gives Elise in the story represents this technique he invented for measuring volume. Elise submerges the ball in the bucket and then watches the water level rise and fall. When the water level goes up, the ball is bigger, when the water level goes down, the ball is smaller, meaning she was getting closer to the wizard.

This brings us to something called Archimedes' Principle, which tells us how and why things float. Basically, when we put an object underwater, we are pushing up some of the water up and out of the way. That water wants to get back down, so it pushes down on the rest of the water, which pushes up on the ball. This is called the buoyancy force.

The more water something displaces, the more buoyancy force there will be on it.

The boat in the story floats because of the force of buoyancy. The weight of water the boat displaces has to be exactly equal to the weight of the boat for it to float. When the queen steps into the boat, the boat will sink a little further into the water. This is because it needs to displace more water for the

buoyancy force to be strong enough to lift both the boat and the queen.

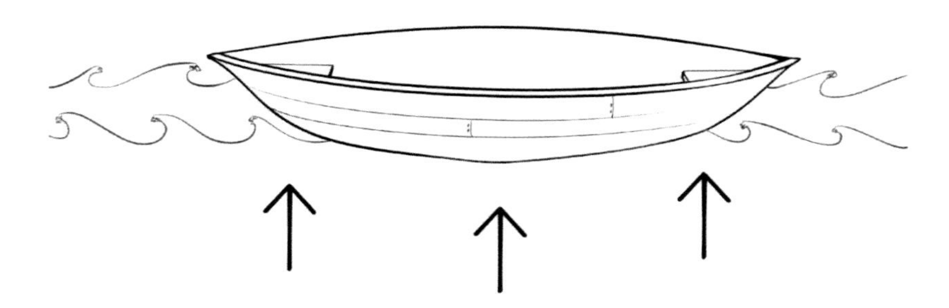

Edmond's other brother in the story, Bernoulli, represents the famous Swiss scientist and mathematician Daniel Bernoulli, who was born in the Netherlands in 1700. Bernoulli is the creator of Bernoulli's Principle, which relates the speed, pressure, and depth of a fluid. It's extremely useful for figuring out how much air or water pressure there will be if you go up higher or down lower in a fluid (or in the air), or if the speed of wind or water will change.

Like Archimedes, Bernoulli loved mathematics, and wanted to study it in school. His father, though, who was also a mathematician, wanted him to study business so that he would earn more money. Bernoulli refused, and they compromised. Bernoulli agreed to study medicine and become a doctor, as long as his father would teach him mathematics.

The magic cup in the story illustrates one feature of Bernoulli's Principle: the deeper you go underwater, the greater the pressure is. So, if you have a container of liquid with holes in the side, the water will come out faster from the lower holes than the upper holes, because there is a greater difference in pressure between the inside and the outside when you're closer to the bottom.

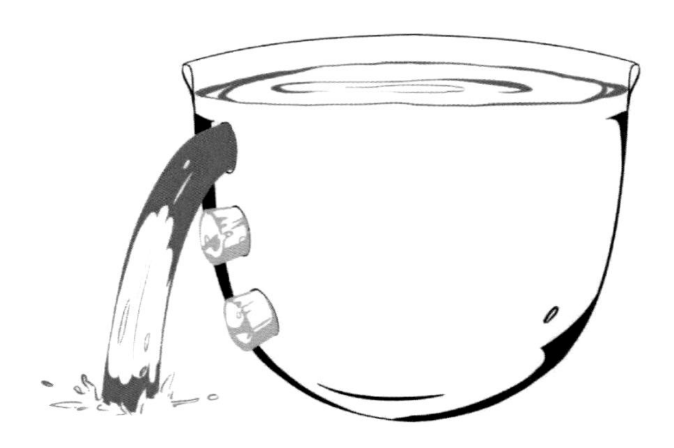

Here, Bernoulli's study of medicine became useful. He found that if you had a pipe with water flowing in it and you wanted to know the pressure of the fluid, you could poke a little hole in it with a straw and see how high the water rose. The higher the pressure in the pipe, the higher the water would rise in the straw.

Horrifyingly, this quickly became used to measure peoples' blood pressure (they would actually poke little glass tubes into people and see how high their blood rose in the tube) and it wasn't until 1896 (over 100 years later) that our current, non-painful, method was invented.

Another idea closely linked with Bernoulli's Principle is the idea of flow rate. When the queen comes to the third island, she has to guide her boat up a canyon that becomes narrower

and narrower. In the canyon, there is a certain amount of water flowing. When the canyon is wide, the water can flow slowly. But when the canyon gets narrower, all that water has to make it through less space, so it has to go faster. (To put it in more science-y terms, the volume of water flowing through any point must be the same.)

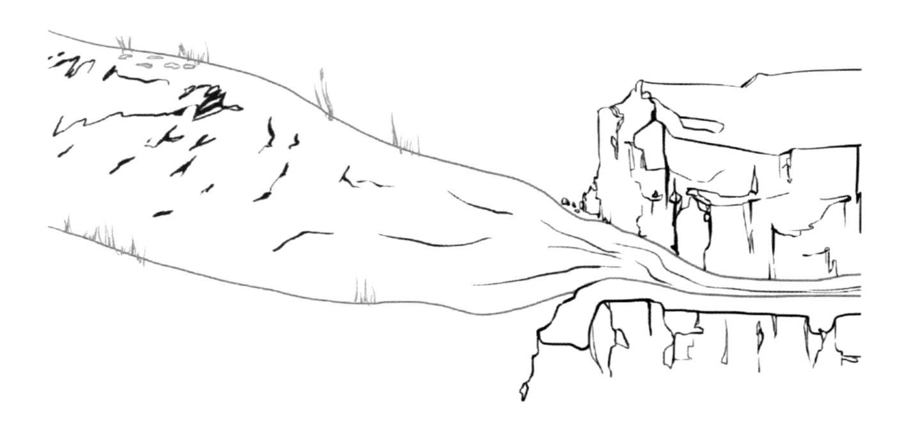

You can experiment with the concept of flow rate yourself. Try putting your thumb over the end of a hose, partially blocking the flow of water. You'll see that the water sprays out much faster. This is because you've given it a smaller space to flow through, so it has to speed up.

When Elise finally catches up with the wizard and he asks her to free his boat from where it is trapped, she goes for a swim, investigating the lake where the boat is.

Elise discovers during her swim that the little lake is connected to the outside water by a tunnel. The tunnel illustrates an interesting thing, which is that the water level in either end of the tunnel would have to be the same, because both ends have to be balanced. The tunnel is a sort of natural manometer, which is a device for measuring pressure. In simplest form, it is a u-shaped tube. If you fill it with water, the water levels will balance on either side. If you connect one end to something with a higher pressure, it will cause the water on the other side to rise (very similar to Bernoulli's blood pressure measuring technique).

Once she discovers there is a tunnel to the outside, she uses something called Archimedes' Screw (another of Archimedes cool inventions!) to empty the water out.

Archimedes' screw is a way of moving water (or sand or grain, lots of things) from one place to another, and it works in an interesting way. Imagine in the diagram above that the screw isn't turning. As long as the seal between the walls and the blades is good, the water won't leak out. Because of the way the loop works, to get to the next lower point, the water would have to go up. It would have to go over the central pole. (The key here is that the screw only works when it's at an angle like this. It wouldn't work at all if it were vertical, because all the water would run out. But, because it's at an angle, the water would have to flow up to get to the next low point.)

Once we can see that the water is trapped, now let's think about what happens as we rotate the screw. As we rotate it, the water gets pushed up the incline. It still stays in the lowest portion of its little space, but the lowest portion moves up the tube as the screw rotates. All you have to do is rotate the screw and the water is lifted through it.

If you would like to experiment with fluid mechanics in your everyday life, here are some things you can try:

You may have seen this one before, it's a classic, but it's really interesting. Take a straw, hold it straight up and down, and stick it halfway underwater like you normally would. Then, put your finger over the top so it's plugging one end of the straw. Then, lift your straw out of the water. You'll notice that the water stays inside the straw. This amazed me when I was a kid, but it amazes me even more now that I understand what's happening. You might think that the water is just stuck inside or something, but really what's happening is that the air under the straw is pushing up on it. Normally, when both ends of the straw are open, the air is pushing on the top part, too, so it balances out, but when we cover the top with our finger, now the air is only pushing on the bottom, and it's powerful enough to hold the water up.

We call this 'atmospheric pressure' and because we're feeling it all the time, we don't notice it, but it's incredibly powerful.

If you're interested in learning more about this, try searching YouTube for the video Space Straw by Vsauce.

To play with more of the invisible but powerful effects of air pressure, try searching for 'vortex cannon project' which tells you how to build a device that will make toroidal (donut-shaped) air pressure missiles. To see a cool illustration of this, find Physics Girl's 'Crazy Pool Vortex' video on YouTube.

Another simple illustration of fluid mechanics that you can do is to get a paper or plastic cup, poke three holes in it at various depths, and then fill it with water. You'll see that the water coming out of the lower holes goes faster. But the distance each stream goes can be surprising. Try experimenting with different depths and differently sized holes.

I hope you enjoyed this exploration of fluid mechanics! If you would like to hear about more cool experiments, YouTube videos, online simulations, and games, as well as get advance notice on new releases, freebies, and discounts on new books coming out, you can sign up for my newsletter on my website: www.MathwithSarah.com. You can also find me on my Facebook page: Math and Physics Fairy Tales.

THE END

References

File:archimedes sphere and cylinder.svg. Wikimedia Commons. (n.d.). Retrieved February 1, 2022, from https://commons.wikimedia.org/wiki/File:Archimedes_sphere_and_cylinder.svg

O'Connor, J. J., & Robertson, E. F. (1999, January). *Archimedes - biography.* Maths History. Retrieved February 1, 2022, from https://mathshistory.st-andrews.ac.uk/Biographies/Archimedes/

Toomer, J. J. (n.d.). *Archimedes.* Encyclopædia Britannica. Retrieved February 1, 2022, from https://www.britannica.com/biography/Archimedes

Wikimedia Foundation. (2022, January 26). *Archimedes.* Wikipedia. Retrieved February 1, 2022, from https://en.wikipedia.org/wiki/Archimedes

Wikimedia Foundation. (2022, January 30). *Daniel Bernoulli.* Wikipedia. Retrieved February 3, 2022, from https://en.wikipedia.org/wiki/Daniel_Bernoulli

Made in the USA
Middletown, DE
12 August 2024

58870393R00029